NATURE'S LANDSCAPES

Grasslands and People

Catherine Horton

WAYLAND

NATURE'S LANDSCAPES

History and geography are closely related. This series looks at six distinctive landscapes of the world, and shows how the nature of his surroundings has governed Man's development and thought, and still deeply influences our way of life today.

Deserts and People

Tundra and People

Rivers and People

Mountains and People

Jungles and People

Grasslands and People

The author would like to thank Ben Burt from the Museum of Mankind for his valuable help in checking the text.

First published in 1982 by
Wayland (Publishers) Ltd
49 Lansdowne Place, Hove
East Sussex BN3 1HF, England

ISBN 0 85340 925 0

Phototypeset by
Direct Image Photosetting, Hove, Sussex.
Printed in Italy by G. Canale & C.S.p.A., Turin
Bound in the U.K. by the Pitman Press, Bath

Contents

1 Grasslands in perspective

Grasslands are often divided into two broad groups: those that occur naturally and those that are man-made. However this distinction can be misleading, for Man has not only created grasslands where they previously did not exist; he has also considerably influenced the character of most of the world's natural grasslands.

To fully appreciate the extent of Man's influence, it is necessary to look first at grasslands as they were before human communities appeared. Evidence suggests that about 16-18 million years ago, changes in the world's climate resulted in the disappearance of many forests. These were often replaced by grasslands, where soil conditions and rainfall favoured their development. During this time too, mammals were spreading over the Earth. Many of them were undergoing important developments in their physical structure; in some cases, changes in limbs and teeth were giving rise to the large hoofed mammals that graze on grasses and other plants.

Neither of these events happened rapidly; they took place gradually, over thousands of years. With them came other changes that eventually resulted in what scientists call the grassland *ecosystem*. An ecosystem is a specialized environment determined by soils, rainfall, temperature, and air. Within this environment there exists a community of plants and animals that are adapted to living in the environment and are dependent on one another for survival. Such an ecosystem is held in a delicate balance and it is maintained only as long as each of its elements continues to perform its special role. Therefore a change in any one of the elements results in some change in the ecosystem. For example, an alteration in the amount of rainfall over a long period has dramatic effects on grassland plants. Where there is too little rain, grassland reverts to desert; where there is too much, trees and shrubs gradually take over from the grasses. Events such as these are natural

Opposite *Two cowboys ride through the ripening corn on their farm in the prairies.*

7

Above *An artist's impression of a prehistoric mammoth hunt on the grasslands.*

and are part of the continual climatic changes that occur on Earth over long periods of time. However when Man interferes with nature these normal processes are interrupted.

Early Man was as much a part of his environment as the other animals with which he shared it. He scavenged for food, gathering plants and their fruit and possibly hunting those animals that were most easily captured without the aid of weapons. When Man learned to control fire and make effective weapons and tools, he had the means to begin to control his environment, and to become a more efficient and effective hunter.

At this stage, Man was primarily a hunter-gatherer, moving from one place to another in search of food. Gradually however, he began to domesticate certain animals and to cultivate certain plants. Mankind now became more diversified: while some peoples continued to lead a nomadic life, others established more permanent settlements where they could raise their crops. The beginning of agriculture marked an important stage in human history.

If we take an enormous leap from these early settlements to the present day, and consider that the majority of the world's grasslands are now used for some form of agri-

culture, we have some indication of the extent of Man's influence on the grasslands. On many grasslands, the rich variety of wild plants and animals has been replaced with a contrastingly restricted number of cultivated crops, or with a few types of domestic animals. Those grasslands where gazelles and zebra can roam freely, feeding on nature's grasses, are now scarce and even these show the marks of Man's interference. In addition, many of our grasslands exist at the expense of other habitats, particularly forests, which have been burned or cut down to make way for agricultural land. In other places, vast areas of grassland have been mismanaged and over-used and are now great unproductive areas of dust and sand.

Man's management of the grassland habitat has not been very impressive. Unfortunately there is little about the future that looks promising if we continue to repeat the mistakes of the past. With the advent of early agriculture, the world population began to rise at an ever-increasing rate. It is rising in even more alarming proportions today. The crucial problem facing Contemporary Man – how to feed the world's peoples – is likely to grow worse in the foreseeable future. It is a problem that concerns the use of land for food production and therefore the proper management of our grasslands.

Above *Today the world's grasslands produce nine-tenths of the world's food.*

9

Above *Starving children in northern Uganda. Proper management of our grasslands is vital if poverty and hunger are to be overcome.*

Right *A scenic view of tropical savannah in Kenya.*

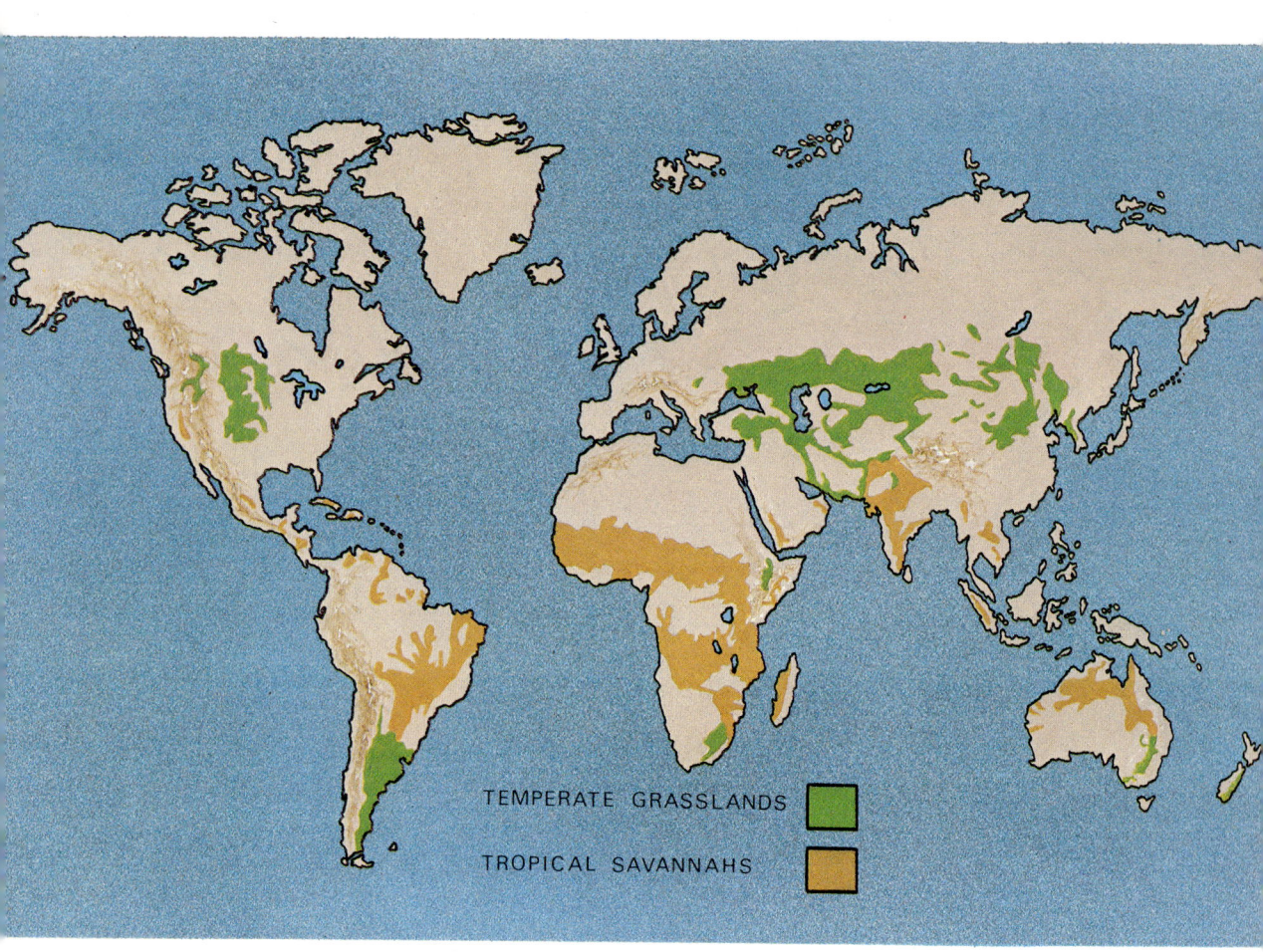

TEMPERATE GRASSLANDS

TROPICAL SAVANNAHS

Above *Temperate or tropical grasslands are found on every inhabited continent.*

2 The grassy plains

Like deserts and forests, grasslands are one of the main vegetation zones of the Earth; altogether, grasslands make up almost a quarter of the total land surface. With the exception of Antarctica, they are found on every continent and the largest lie on great plains in the interior of these landmasses. These are the major grasslands of the world with which this book is primarily concerned. Smaller areas of permanent grassland also occur throughout the world and include alpine meadows, cliff-top grasslands near the sea and old pastureland that has been maintained for many years by Man and domestic animals.

Just as a forest can be defined as a place where trees are the main type of vegetation, so a grassland can be defined as a region where grasses form the dominant type of plant life. Yet grasslands vary enormously from one part of the world to another and they even show some variation within themselves. For example, one area of a grassland may be almost pure grass, but other areas may be a mixture of grasses and other plants, including trees and shrubs. This is because the type of plants that grow in any area depend mainly on what the climate and soil are like, and these conditions vary both locally and world-wide.

As climate plays a particularly important role in the formation and continuance of grasslands, it is generally used as a basis for classifying the world's grasslands into two broad groups: those that occur in *temperate* regions and those that occur in *tropical* regions.

Temperate grasslands

Temperate grasslands usually have dry, warm summers and cold winters when much of the land may be covered with frost or snow. Rain falls mainly in spring or early summer and the annual amount of rainfall is rather low—between 25 and 75 centimetres (10-30 in.) a year.

Extremes in temperature, lack of moisture and frequent winds discourage the growth of trees and shrubs. Temperate grasslands therefore tend to be flat or undulating, treeless areas. Three of the largest and best-known types are the North American prairies, the South American pampas and the Eurasian steppes.

The North American prairies are among the most ancient of the world's grasslands. They lie in the central and western part of the continent and include a large part of the area that is called the Great or High Plain. To the north and east they border on forest and woodland and to the west they extend to the foothills of the Rocky Mountains. Southwards, the prairies reach down into the semi-arid scrubland of New Mexico and Texas.

The climate on the prairies varies from region to region, but generally it shows seasonal extremes. In winter, Arctic winds bring cold air from the north, while warm air travelling northwards from the Gulf of Mexico results in hot and

Below *Unending fields of wheat stretch as far as the eye can see across the North American prairies.*

often dry summers. These winds from the south bring most of the year's rain in spring, but the amount of rain that will fall in any one year is largely unpredictable. On the Great Plains, for example, a year of wetness may be followed by a year of drought. Rainfall also varies east to west across the prairies, with less and less rain as the grasslands near the Rocky Mountains. This affects the type of vegetation that is present at different longitudes, so that the prairies can be divided into three zones: tallgrass prairie, shortgrass prairie and a zone of mixed grass prairie between the other two. The names of these zones refer to the characteristics of different grass species, but they also indicate the presence or absence of trees and shrubs. On the moist, rich soils of tallgrass prairie, trees grow along rivers and streams, but the shortgrass prairie, with its sandier soils, is almost treeless.

Today, most of the prairies have the uniform appearance that often results from cultivation, for much of the natural grass cover has been replaced with agricultural crops or is used as pasturage for livestock. Yet there are still a few

Above *Temperate grasslands can become very cold in winter. An icy view of the Eurasian steppe.*

15

small areas where a variety of prairie grasses and other flowering plants grow as they have for thousands of years.

The vast continental plain that stretches from eastern Europe across Asia as far as Siberia is known as the Eurasian steppe. (The word *steppe*, which means plain, is also sometimes used for other dry, temperate grasslands.) In many ways it is similar to the North American prairie; spring rains encourage the growth of grasses and other plants, summer is hot and very dry, and the winter harsh. However, summer is shorter on the steppes and there is no gradual transition to winter, which comes suddenly and is the longest season of the year.

The central part of the steppe is typical, treeless grassland but towards the northern forest the open plains alternate with wooded areas on what is called the forest steppe. The southern regions fringe the central Asian deserts, and here the steppe is dry semi-desert. Soils are mainly sandy but clays are present in some places. The clay soils contain more nutrients and can therefore support a richer plant life. In eastern Europe almost all the plain is now agricultural land, but in Asia there are areas that still have a more natural vegetation.

The pampas is the large temperate grassland of South America. It occupies an enormous area in the south-eastern part of the continent and makes up much of the agricultural land of Argentina, Uruguay and southern Brazil. Originally, native grasses covered the rich, fine-grained soils; they included the tall, plumed pampas grass that grew up to three metres in height. The climate on the pampas is more moderate than it is on the steppe but there are still marked differences between winter, which comes in July, and summer, which arrives in January.

The annual amount of rainfall varies from east to west, with the wettest areas near the coast and the driest towards the interior. Most of the rain falls in summer.

Temperate grasslands also occur in Africa—the South African veldt—and in south-eastern Australia. In Australia particularly, much of the land has been taken over for agricultural use and sown with cultivated grasses.

Opposite *Gauchos in brightly coloured dress herd sheep on the South American pampas.*

16

Savannahs

Below *Elephants at a waterhole on the African savannah.*

Tropical grasslands or *savannahs* occur in hot climates where there is little variation in temperature from one season to another. The year is divided into periods of rain, which may come once or twice a year, followed by periods

of drought. Rainfall is greater than on temperate grasslands and often exceeds 100 centimetres (40 in.) a year, but it varies from region to region so that areas fringing on desert will have less than those that lie close to tropical forests. Major tropical grasslands are found in Australia, Africa and South America. In South America, those in the central and southern regions of the continent are known as *campos,* and those in the north, as *llanos*.

The largest African savannahs lie north and south of the equator, surrounding the tropical forest in the centre of the continent. As on other tropical grasslands, trees and shrubs are scattered over the plain or grow close together in river valleys and other places where water is plentiful. The African savannahs are often regarded as the most spectacular of all the grasslands. This is largely because many of them still retain an abundance and variety of their original plant and animal life.

On the South American campos, winters are warm and summers very hot, and it is in the heat of summer that most of the rain falls during heavy thunderstorms. As on most tropical grasslands, the vegetation is dry and scorched during the dry season, but when the rains come the land has a green and lush appearance.

Artificial grasslands

Although many of the grasslands discussed previously have been maintained by Man for thousands of years, they are natural in the sense that they were originally formed by natural processes. Artificial grasslands, however, are wholly man-made and have been created in areas formerly occupied by other habitats. Good examples of this type are the temperate grasslands of Britain and New Zealand and the tropical grasslands of India, all of which are the result of extensive forest clearance. Grasslands have also been made by draining marshland and irrigating deserts. These methods have been used throughout the world to meet the need for more and more agricultural land.

Below Grasslands in Britain have been artificially created by clearing the forest that once used to grow on these Devon hills.

3 The plant community

Grasses are herbaceous (non-woody) plants. They include the familiar grasses of roadsides and meadows, the more cultivated varieties sown as lawns and a great number of other grasses, some highly specialized. Altogether there are some 8000 species, some of which grow to over 4 metres (13 ft.) and more in height. Yet despite their great number and variety, they all have one thing in common: they are remarkably well adapted to survive even the most adverse conditions.

Grasses have slender, usually hollow stems and narrow leaves. Their very small flowers grow in spikelets which are sometimes grouped together to form a flower head. Although the individual, petal-less flowers are barely noticeable, the flowering heads are often the most conspicuous and largest part of the plant. As producers of seeds they are also one of the most important, for in many cases seeds ensure the continuous survival of the grass species.

Below *Long slender stems characterize this red oat grass.*

All grasses grow quickly, but many of them mature, produce seeds and die within a year. These grasses are called annuals. Annuals that grow where water is scarce may complete their life cycle in one short season. For example, on desert fringes most of the grasses are annuals. They produce a green carpet for only a few weeks when the rains fall; then they shrivel and dry during the drought. Annual grasses are sometimes called pioneer grasses because they can often grow on soil that is too poor for other types of grass. When they die, their roots, leaves and stems decay and add valuable *nutrients* to the soil. The soil is gradually enriched so that more permanent plants can take over from the annual grasses.

Grasses which continue to grow and reproduce from one year to the next are called perennials. Although they may also produce seeds, some of them are able to propagate by producing new side shoots from *stolens* (stems that creep

Above *Tussock grass grows in thick clumps and is well adapted for survival in harsh conditions.*

21

along the ground) and *rhizomes* (stems that grow under the ground). The new shoots may grow so close together that they form large clumps, or they may interlock and form thick turf. Matted turf is advantageous both to the soil and the plant, because it traps rainwater near the surface so that water is readily available for the roots. It also protects the soil against erosion. In addition, extensive root systems help to retain moisture in the soil.

The ability to produce seeds rapidly and propagate by several different methods enable some grasses to survive in a demanding environment. Fire, drought, high winds, and freezing temperatures are frequent occurrences on many grasslands, but because of their special qualities the grasses are able to withstand most of them. Rhizomes, protected under the soil, allow the plant to lie dormant during a period of drought or extreme cold. They can then put up new shoots when warmer weather comes or the rains arrive. In some grasses the seeds are also able to tolerate unfavourable conditions. For example, they may be able to germinate after long periods of dryness or even after they have passed through the gut of an animal.

Animals sometimes have damaging effects on grasses. They trample the plants underfoot and eat both the top-growth and the roots. However, because grasses grow from the base of the stem rather than the top, they can, in many cases, continue to grow after they have been grazed by wild animals. In addition, some grasses contain a hard substance called silica which reduces damage to the plant from animals' teeth.

The other members of the grassland plant community form two large groups: non-woody herbaceous plants other than grasses (sometimes referred to as *forbs*), and trees and shrubs. Most are specially adapted to the grassland habitat and share some of the characteristics of the grasses, such as the ability to grow and spread quickly.

When the rains come, many wild flowers burst into bloom. On the Eurasian steppes there are bulbous plants such as lilies, tulips and anemones, as well as daisies and

salvias. Lilies are also found on the savannahs of Africa, as are many members of the pea family, such as clovers and vetches.

These are not the only flowering species however. Trees and shrubs may produce delicate or flamboyant flowers in spring. Chokecherry and wild plum are among the flowering shrubs of the North American prairies and flowering acacias are common on the grasslands of Africa, Australia and South America.

The acacia trees and shrubs usually flower just before the rainy season. As the moisture-laden clouds gather, the increased humidity of the air causes the buds to open. They can then be pollinated by wind or insects before the heavy rainfall damages their blossoms. Acacias are members of

When the rains come, the African savannah becomes green and lush.

23

Below *Wild flowers mingle with grass during spring on the South Downs in England.*

the pea family and their fruit is contained in a pea-like pod. They have feathery leaves and many of them have long, hooked thorns which give their branches some protection against browsing animals.

The baobab is another tree found on tropical African and Australian grasslands in areas of low rainfall. The trunk

and lower branches are composed of a spongy, cork-like material which enables the tree to conserve large quantities of water. Like the acacias, the baobab sheds its leaves annually, but in this case the tree bears its leaves for only a few weeks. Because it is leafless for most of the year, there is no danger that any of the water stored in the baobab will be lost through the leaves.

On natural grasslands the climatic pattern generally produces dramatic differences in vegetation from one season to the next. On grasslands that have been created by Man the situation is often very different. This applies particularly to grasslands that have resulted from the clearing of forests in areas that have a high rainfall spread out over most of the year. Here the grasses and other plants are able to flourish for longer periods, with different species flowering in different seasons.

Above *The baobab tree bears its leaves for only a few weeks every year.*

25

Above *Wildebeeste, warthogs and zebra gather at an African waterhole.*

4 The animal community

Large hoofed animals, such as antelope, are often the most conspicuous of the grassland animals, yet they are only a part of the varied and complex animal community. Within this community each member has a distinctive role, or niche, in the grassland ecosystem. The niche an animal occupies is usually defined by the type of food that it eats. Food provides animals with energy necessary for life and comes from two sources: plants and other animals. Animals that eat plants are known as *herbivores*, while those that eat flesh are called *carnivores*. A third group, the *omnivores*, eat both plants and animals. Yet whatever type of food an animal eats, it is dependent in some way on the other animals it shares its habitat with, and on the habitat itself.

The way in which plants and animals depend on one another can be seen in a simplified version of what scientists call the food chain. The food chain starts with green plants, which provide food for herbivores. The herbivores in turn are a source of food for carnivores. Omnivorous animals take part in the food chain at two stages, and like many flesh-eating animals, they may themselves be consumed by other carnivores. The amount of plant food that is available on a grassland will determine how many herbivores there are, for if food is scarce the plant-eaters will either die of starvation or will move to areas where food is more plentiful. In the same way, the number of herbivores controls the number of carnivores. This system operates in all plant and animal communities.

The herbivores

Many of the grassland herbivores are too small or too well hidden to be seen by a passer-by, for they live among the vegetation and even beneath the soil. Some small insects and other invertebrates (animals without a backbone) live

Left *Harvester ants carry grass seeds to their underground nests.*

Below *Something has aroused the curiosity of this marmot from the prairies.*

permanently below ground, feeding off plant roots; others, such as the harvester ants of Africa, live in underground colonies but come up to forage for leaves and seeds. While many insects–aphids for example–can cause enormous damage to plants, others may have a beneficial effect. The harvester ants' underground life helps to mix and aerate the soil as well as to fertilize it, just as earthworms and burrowing mammals do in other areas.

Small mammals are plentiful on grasslands, especially the group known as rodents. Rodents' teeth are specially adapted for gnawing through tough plant tissue, which enables them to feed on every part of a plant–from roots to hard seeds and fruit. Many of these animals are burrowers, living in underground tunnels and chambers. Burrowing rodents include the tucotuco and mara of the South American pampas, the suslik and the bobac marmot of the Eurasian steppes and the gophers and prairie dogs of North America.

Rabbits and hares are not true rodents but they share some of the rodents' characteristics. In large numbers, rabbits can become pests, destroying vegetation and thereby depriving other plant-eaters of food and dwelling places.

Although many birds feed on grassland vegetation, most of them also take insects and other small animals. Exceptions are the burrowing parrot of South America and prairie chickens of North America. The burrowing parrot makes long tunnels in the earth with a chamber at the end for its nest; it lives on a diet of flower buds, seeds and roots. The sage grouse, a type of prairie chicken, lives almost exclusively on the leaves and shoots of sage-brush.

Most large grassland herbivores are mammals, and the majority of these are ungulates–herbivorous mammals with hooves. Like the smaller animals they are specially adapted to life on savannahs and plains. Antelopes and deer, for example, have long legs which enable them to run swiftly away from predators, and individuals are also protected by living with others in large herds. Powerfully-built

Below *Large herds of zebra can be found grazing on the African savannah.*

Above *This giraffe thoughtfully munches an acacia leaf.*

herbivores such as the elephant and rhinoceros are protected from most hunters by their sheer size–although strong tusks and massive horns can be used as weapons if necessary.

Ungulates have specialized mouthparts and teeth to deal with various types of plants. Some have delicate and refined mouthparts with which to select the tender parts of grasses and herbs; they then clip off the stems or leaves with their sharp incisor teeth. Molar teeth are adapted for grinding tough plants. Some browsers, such as the giraffe, have long tongues that enable them to grasp leaves that grow between the thorns of acacias. Grazing by wild animals helps to maintain grasslands, for continual clipping and cutting stimulates the growth of grasses.

The large herbivores are unevenly distributed over the world's grasslands. The African savannahs support the richest populations which are scattered over every type of grassland habitat, from dry semi-desert to wooded savannah. They include gazelles, elands, zebra, white and black rhinoceros, elephants, impalas, and gnus or wilde-beest. Wallabies and kangaroos are the only large wild herbivores of Australian grasslands, and in South America this group of animals is represented solely by the pampas deer, which is now rare. Enormous herds of buffalo (American bison) and pronghorn antelopes were once present on the prairies, but due to persecution by Man, they are now found only in small numbers on nature reserves. The same thing almost happened to the saiga antelope of the Eurasian steppes, but fortunately they were able to recover and are now relatively abundant in certain places.

Predators and scavengers

Just as herbivores help to maintain grasslands by con-tinually cropping plants and thereby ensuring their growth, so carnivores have a beneficial role in the animal community. Part of this role is to keep the animal popu-

lation within certain limits, so that there is less danger of overcrowding. Large predators such as the lion, which preys on a variety of antelope, also help to keep species healthy. By taking mainly the weak, the sick and the older members of a herd the young and fit members are left to breed.

Large mammalian predators include members of the cat family, and the wild dogs. Of the wide variety of birds that inhabit grasslands, some are fierce predators. The bataleur eagle of Africa hunts snakes, birds and small rodents; the prairie falcon of North America and the saker falcon of Eurasia feed on birds and some mammals. While birds are

Above Kangaroos are the largest herbivores on the Australian grasslands.

numerous on grasslands, the number of reptile species is relatively few. Of those that are present, the most vicious are some species of snake. The puff adder is sometimes found on African savannahs and its relative the rattlesnake occurs on the prairies. Both snakes kill their victims by injecting them with poison from fangs located in the upper jaw.

Scavengers are animals that feed on dead flesh and the remains of kills left by other animals. They include the jackal, vultures and vulture-like birds such as the marabou stork, and some beetles. All scavengers perform a valuable service by ridding grasslands of decaying matter which might otherwise spread disease.

Below *A pride of lions, including some very young cubs, feeds together on a kill.*

This rattlesnake is coiled ready to strike. 33

5 Early peoples

The history of grassland peoples and the origins of Man appear to be closely linked together. Although scientists are still debating the details of how and where Man first evolved, it is known that from early times he lived on the plains of Africa. Archaeologists have found evidence in East Africa that shows that Apeman (*Australopithecus*) existed there some five million years ago, and that Upright Man (*Homo erectus*) lived there between one and a half million and three hundred thousand years ago. Although some scientists believe that Man originated from Africa, fossil records indicate that Early Man also inhabited parts of Europe and Asia.

Modern Man (*Homo sapiens*) emerged about 40,000 years ago. At an early stage in his history he lived on open grasslands and the edges of forests. Before this time, however, Early Man had learned to control and make fire and had learned to use and make tools and weapons. Prior to these developments food sources would have been limited to carrion, small creatures and wild plants. Even after Man had learned to use his own roughly-made tools,

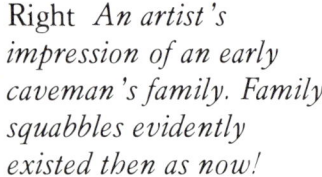

Right *An artist's impression of an early caveman's family. Family squabbles evidently existed then as now!*

his weapon-making was not far enough advanced to enable him to kill big game.

Gradually Man became skilled at making tools and weapons and at hunting; however it is probable that a diet of meat was supplemented by plant food such as fruits, seeds and roots. In some cases, vegetation may have made up 75 per cent of the diet. Fish and shellfish were a further source of food when they were available, and widened the number of resources that could be used when other food supplies were scarce. The knowledge that Early Man both hunted and collected wild plants has given rise to the idea that our ancestors were hunter-gatherers, rather than just hunters. Only in exceptional cases, in both ancient and recent times, has Man relied exclusively on hunting.

Above *The remains of an early settlement called Cisbury Ring in southern England. Grassland hilltops were frequently fortified during troubled times.*

35

Hunters and gatherers

Hunting and gathering appears to have been well established as a way of life between a million and half a million years ago. By studying ancient sites and the few hunting and gathering societies that exist today, *anthropologists* and *archaeologists* have been able to put together a picture of how the early hunter-gatherers probably lived. They led a nomadic life, travelling from place to place in search of food, and because of this they would have had few possessions. Several families may have travelled together in small bands, and although some hunting may have been done by individuals, it was probably necessary to form hunting parties–for example when driving or 'beating' animals through bush and tall grass. Because they lived and worked together, these peoples must have had some basic form of social organization that enabled them to share food resources, rear children and care for the sick, as well as hunt.

Below *Early Man probably had a considerable understanding of plants. This scene shows an early attempt at bread making.*

Hunter-gatherers probably had considerable knowledge and understanding of plants and animals, and may even have developed a close relationship with them in their religious beliefs. In addition, they may have deliberately encouraged the growth of certain plants that they found to be most productive and useful, transplanting them to places where conditions were more favourable for their growth. It is known that they used fire as a means of attracting game and also for driving animals into the open. Setting fire to bush and tall grass improved their chances of bringing an animal down by increasing the hunters' visibility. It also promoted the growth and spread of grasses into wooded areas, therefore increasing the pasturage for wild grazing animals.

During the successive ice ages (which began some 2½ million years ago) vast regions of Africa and Europe that later became forest or desert were open, grassy plains. The Sahara, for example, had fertile areas that were inhabited by hunter-gatherers. While much of Europe was covered with glaciers, areas in the south, such as France and Spain, had regions of steppe vegetation that were rich in animal life. These areas were inhabited by nomadic hunter-gatherers who left behind them fine examples of their art in the remarkable cave paintings of south-western France and northern Spain. These paintings tell us something of the animals that were hunted and of Man's relationship with his environment. Most of the cave paintings are of animals – horses, bison, deer, and bulls – but there are also outlines and prints of human hands. A few show human beings, but in this case they are sketchily painted, with featureless faces, whereas the animals are done in more detail. Small objects have also been found in the caves, such as beautifully engraved antlers.

Scientists disagree as to the purpose of the cave paintings and there is not enough evidence to indicate their meaning. One theory is that they were painted before a hunt, to bring success to the expedition. Another idea suggests that Early Man was trying to depict the spirits of the animals that he had to kill in order to survive, believing

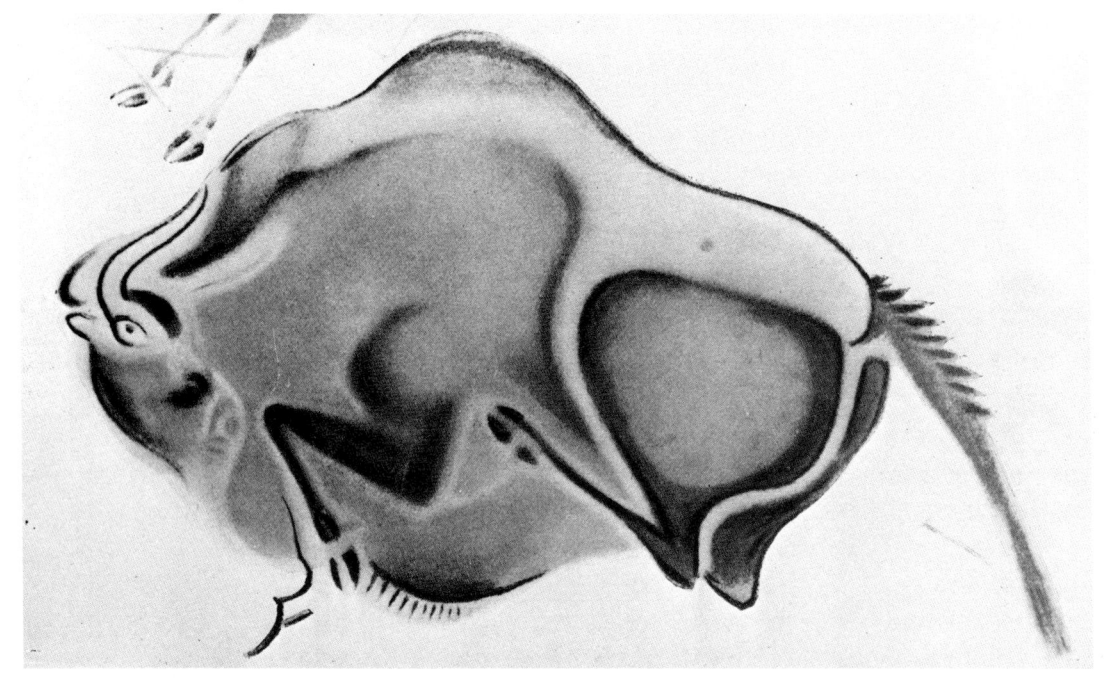

that the animals' spirits would live on in the paintings.

These cave paintings in Europe were probably done between 35,000 and 10,000 years ago. In Africa similar works of art survive on rocks but because they are more exposed they have not been as well preserved. One of the finest examples comes from Tanzania, and shows both human figures and a variety of animals. Rock paintings are also found in regions of the Sahara.

The spread of Man

The continual need to search for food may have been responsible for the spread of Man throughout Europe, Asia and later, Australia and the Americas. Yet whatever the reason, hunter-gatherers did undertake these migrations. Looking at a map of the modern world it is difficult to see how these early peoples could have crossed the Bering Strait and more incredibly the Timor Sea into countries that until then had been empty of humans. The explanation appears to be that the huge ice sheets that

Above left This painting of a buffalo was drawn on the wall of a cave in Spain.

Below left Hunting was probably carried out in bands of two or three families.

Below Northern people were very dependent on the reindeer for both food and clothing.

formed during the ice ages drew up water from the oceans, causing a considerable drop in sea level throughout the world. As the sea level dropped, more land was exposed. This created a land bridge between the north-eastern tip of Asia and what is now Alaska, and it also narrowed the Timor Sea. The journey from Asia to America (some 84 kilometres) was made on foot, but that to Australia could only have been accomplished by boat. It meant travelling a distance of some 96 kilometres over open water. In both cases the journey must have been made several times.

Evidence shows that Man was relatively widespread in Australia 18,000 years ago, and was present in what is now California 26,000 years ago. The first crossing to America however was made as early as 40,000 years ago, and the last (which brought the peoples known as Eskimos) some 10,000 years ago.

Herders and farmers

The hunting and gathering way of life lasted about two million years. It was gradually replaced by one that was to have enormous consequences for mankind. In view of its importance, some experts term the change from hunting and gathering to cultivating crops and herding as the 'agricultural revolution'. The term revolution, however, carries with it a sense of immediacy, but this was not the case. The change to agriculture was not a sudden one, but must have taken place over a fairly long period, with small but significant changes gradually leading to the cultivation of plants and the domestication of animals. Even then, some peoples continued to pursue the hunting and gathering way of life and a few peoples in forests, deserts and Arctic areas still lived by these means until very recently.

The domestication of animals and herding by Man is believed to have happened before the cultivation of plants. Precisely where and when animals were first domesticated is not known, but it was either in Asia or Europe.

Opposite *Pastoralism exists among a number of tribes in East Africa. The herdsmen have enormous pride in their cattle.*

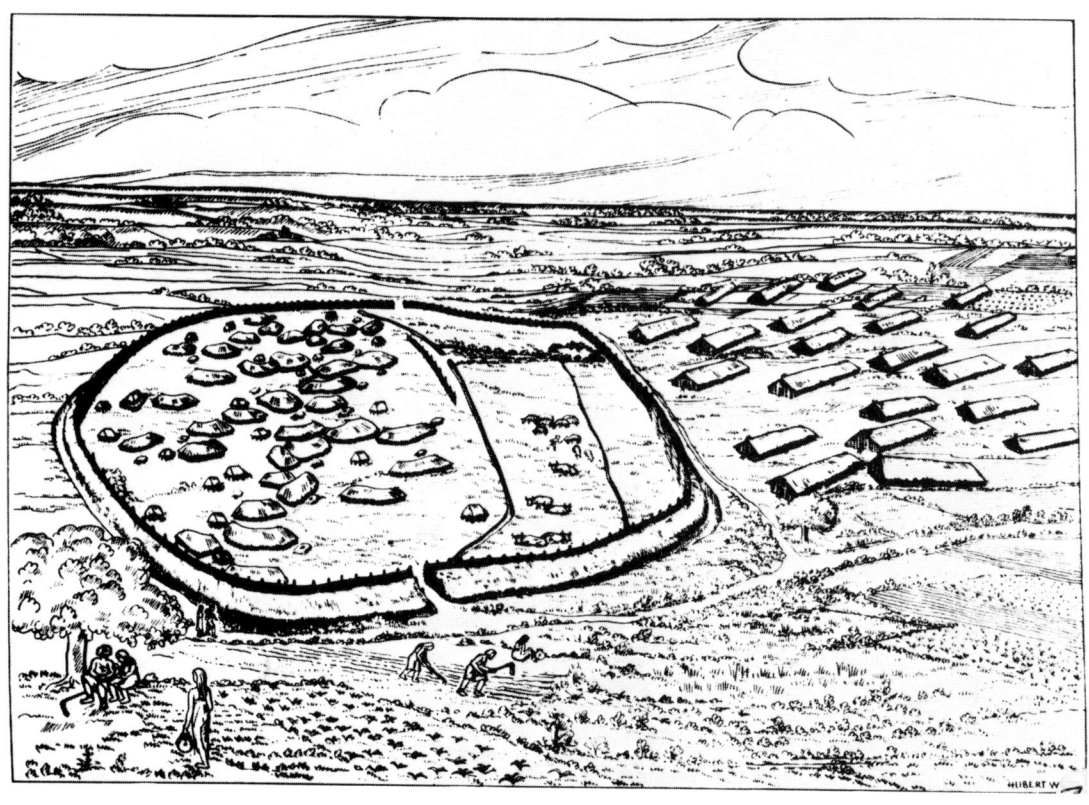

Above *The cultivation of crops resulted in the first semi-permanent settlements being set up.*

Domesticated species were introduced into Africa, but it is probable that at a later date successful attempts at domestication occured there independently.

The first animals to be domesticated were sheep and goats; cattle, horses and camels were tamed and herded later. Herding these animals gave Man a more assured and productive food supply, although he still had to move to and from watering places and fresh pastures. This new way of life—called *pastoralism*—would no doubt have also involved the gathering of wild plants and perhaps attempts at crop-raising, but the dominant concern of the shepherd was with his herds.

Pastoralism, which still exists today in some parts of the world, gave Man more control over his environment and he therefore had a greater effect on his surroundings. As the use of domesticated animals spread, there was an ever-increasing demand for pasture which was met by an increase in land clearance with the aid of fire and the axe. Also, as breeding of livestock continued (mainly by selective slaughtering) domesticated animals became further and further removed from their wild ancestors and finally gave rise to new species and races of animals.

Agriculture, including both herding and the cultivation of crops, developed and spread between 10,000 and 2,000 years ago. Most scientists agree that it arose independently in different areas at different times, but also spread out from each area into the surrounding regions.

The most well-known, and possibly the earliest centre of agricultural development was in south-west Asia (the Near East) on steppeland surrounding the Tigris and Euphrates rivers. By 10,000 years ago agriculture was well established in this region. Here there was a natural abundance of food, both plant and animal, and because of this the area is sometimes called the fertile crescent. The concentration of wild herbivores and plants, such as the wild ancestors of wheat and barley, may have attracted hunter-gatherers to settle in the area, first to harvest plants and later to cultivate them.

The cultivation of food crops demands a more settled way of life than pastoralism. In south-west Asia it gave rise to permanent communities. This may have been a result of having to care for crops and land, and also having to store the produce of cultivation. It was accompanied by advances in craft skills and an increase in trade. In addition, as the settled way of life became permanently established, certain families would have acquired more power than others, probably by the accumulation of wealth (in the form of surplus crops and other commodities that could be traded), gradually producing a ruling group. These members of the community would probably have directed the distribution of goods and organized the people who specialized in making pottery, tools, cloth and other items.

The events which occurred in the fertile crescent eventually gave rise to towns and cities. However, the development of agriculture did not always produce these results; there were simple farming communities that

Above *Oats, barley, wheat and maize are four of Man's most important crops.*

lacked the division of labour, craft specialization and other features which were characteristic of the south-west Asian communities.

Within 4000 years of the early settlements in south-west Asia, agriculture had spread as far as western Europe. One of the earliest forms of agriculture here (which is still followed in many forested areas in the tropics) was *shifting cultivation*. The method used was to clear land of forest by cutting down the trees and burning, before sowing wheat and barley in the ashes. After some ten to fifteen years of cultivation the soil would become exhausted. The farmers would then move on to another area and repeat the process, leaving the previously-cultivated land to revert to forest. As the population in Europe rose, and more land was cleared, permanent communities were established here, and the forest in most parts was destroyed forever.

Agriculture is thought to have arisen in north-east China between 7000 and 5000 years ago, and there is some question about whether it developed in south-east Asia even before the fertile crescent was cultivated. In Mexico and parts of South America there were cultivators 5000 years ago.

Above *A Sioux Indian encampment.*

6 The Plains Indians of North America

The land bridge that enabled Early Man to pass from Asia to the Americas had also allowed the passage of animals both to and from the two continents. Like the early peoples, the animals had probably spread southwards during those periods when the great ice fields receded. Before the end of the last ice age (approximately 10,000 years ago) the prairies of North America were inhabited by a wealth of animal life. Some of them, such as horses and camels, had evolved on the American continents; others — musk oxen, deer, bison, and mammoths — had crossed over from Asia. These animals, and others such as giant sloths, were successfully hunted by early peoples, providing them with food and skins for clothing and possibly shelter. Archaeological finds of finely-crafted tools, including blades and knives, indicate that these peoples made use of bone and antler as well as wood. Their weapons were the lance and the javelin and later, the bow and arrow.

For several thousand years the early plains people survived by hunting. About 8000 years ago, however, much of the big game began to decline, both in species and numbers. Although the horse, for example, had by now made its way into Asia, it became extinct in America, as did the camels, giant mammoths, sloths, and the sabre-toothed cat. Although the reasons for these extinctions are still a mystery, they were dramatic enough to change the way of life of many of the early plains people. Some game still remained; it was not a complete extinction of animal species but it was serious enough to result in a change from a predominantly hunting life-style to one that involved giving more emphasis to plant foods.

In many parts of the Americas hunting and gathering gradually gave rise to crop cultivation (which in North America had originated in Mexico) and a more settled way of life. This happened to some extent on the prairies, particularly among the peoples who lived along the Mississippi and Missouri rivers. Early crops included beans and squash and the native corn, maize, as well as wild rice in the more easterly parts of the prairies. In other areas, especially on the Great Plains, the peoples continued to lead the nomadic existence of hunters and gatherers.

Thousands of years were yet to pass before the cultural traditions that we associate with modern Indian peoples were to become established. After the big game extinctions and before the Europeans arrived, many modifications and changes were to take place. These resulted finally in the emergence of specialized and complex societies, among them those of the Plains Indians.

The Plains tradition

The recent past on the prairies, as well as elsewhere in America, is usually divided into the time prior to the arrival of the Europeans and the time after they arrived. This is because the invasion of the New World by outsiders eventually had a devastating effect on the peoples.

The Indian people who lived on the prairies before the Europeans arrived were spread across the grasslands in many different tribal areas. There was no agriculture on the western plains and none of the other Indians relied solely on agriculture for their food, although some, such as the Caddoan-speaking tribes of the south, were predominantly cultivators. The people in the east left their crops once or twice a year to hunt the buffalo, a descendant of the larger, ancient bison that had been hunted by the early peoples. There was also a small but important number of tribes that relied almost completely on hunting the buffalo, although deer, antelope and elk were also killed and wild plants were gathered. These people followed the buffalo during its seasonal migrations from winter to summer pastures. They travelled and hunted on foot. Dogs were used to pull the *travois*–a type of sledge–to which all an Indians' possessions were tied. Until the 1700s dogs were the only domestic animals used by the hunters.

The nomadic hunters depended on the buffalo for food, for hides for clothing and shelter, for bones used to make tools and weapons and for dung, which was burned as fuel. Their life was no doubt hard: it was also dangerous. An adult male buffalo was a massive beast; it stood some 180 centimetres high at the shoulders and weighed up to 1350 kilograms. Hunting such creatures on foot took enormous courage and skill.

During the winter, a tribe (all those peoples who lived in peace and regularly associated with one another) would be dispersed in small bands, usually made up of several related families or the followers of a certain leader, or chief. At this time of year the game was dispersed too, and food was relatively scarce. In winter the buffalo was either hunted by individuals or by three or four men. In the summer, when the grass was more abundant and the buffalo gathered to breed, all the peoples of the tribe came together. Large groups of hunters drove the buffalo into wooden compounds, or stampeded them over a cliff to their death.

Below *The Indians depended on the buffalo for food, clothes and tools. Today only a few of these animals remain.*

To hunt buffalo successfully, the Indians had to rely on their knowledge of the herds' migratory habits so that they could intercept them at the right time and in the right place. If for any reason–such as a change in the normal weather pattern–the animals deviated from their usual route, the hunters and their families would be in danger of starvation. It was probably the precarious nature of such an existence that stopped some tribes from hunting as a full-time activity. However the arrival of the Spaniards in the early 1600s was to change this. It was the Spaniards who re-introduced the horse into North America, and by raiding Spanish colonies, rounding up escaped horses and later by trading, the Indians took possession of all the horses they could get.

One of the most important effects of this on the Indians' way of life was that it enabled them to pursue the buffalo with considerable ease and with much greater success. By the late 1700s all of the Plains Indians had horses, and the original tribes who had hunted the buffalo for years had been joined by others who had given up a more sedentary life as cultivators.

The main buffalo-hunting peoples included the Sioux, Blackfoot, Plains Cree, Assiniboine, Crow, Cheyenne, and

Below *The introduction of the horse to America by the Spanish enabled the Indians to pursue the buffalo with much greater ease.*

Comanche. Although each tribe had traditions and practices that were peculiar to itself, they also had much in common. It is therefore possible to describe generally their way of life.

Clothing and shelter

Skins for clothing and shelter were initially treated in the same way; they were stretched onto pegs in the ground or laced on a frame and then scraped with a bone 'flesher' to remove any remains of flesh. Further dressing depended on what the skin was to be used for. Soles for moccasins, for example, were often made of rawhide–the tough, hard-wearing skin that was produced after stretching and scraping. For clothes, bedding, *tipi* covers and other items, the hide had to be made pliable and soft, and this process included soaking the skin in water and rubbing it with a dressing of brains and other animal substances.

Both men and women wore moccasins and leggings; women's leggings reached to the knee, and men's to the hips. The body was covered by a robe, under which women might wear a long dress, and men a shirt. The best

Below *An Indian girl poses in front of her painted* tipi.

garments were usually made of deer or elk skin, but buffalo skin was always used for the warm robe. All hides for clothing were sewn together with buffalo sinew, which was pushed through holes made in the hides with an awl (a pointed, hole-making tool).

Tents, or tipis, were made from 11 to 22 buffalo skins, depending on the size of the dwelling. These were used to cover the wooden poles, usually from pine trees, that were arranged in a circle and brought together at the top to form a cone shape. A hole was left in the top to allow smoke from the fire to escape and the inside walls were often lined with skins as insulation. The entrance to the tipi always faced east, towards the sunrise.

Both clothing and tipis were painted and decorated. Dyed porcupine quills, and later trading beads, were sewn onto clothing in geometric patterns. Clothing for women might have fringes, and elks' teeth decorations; hair or fur fringes were used on men's clothing. Painted designs were made with natural pigments and a bone tool; on men's robes and tipis they were usually realistic drawings depicting special themes. The skin bags and cases for carrying possessions were also decorated by these methods in geometric patterns made by the women.

Religion

The Plains Indians believed that supernatural powers were present in all natural phenomena and in all the living creatures of their environment. They also believed that they could gain such spiritual powers by undergoing certain rituals. One of these was the 'vision quest' which was undertaken by young men. After going off alone to a deserted place they fasted for several days; lack of food and water produced hallucinations, during which the youth believed he received a blessing from the spiritual powers and perhaps spiritual and ritual knowledge.

Symbols of the supernatural were kept in sacred bundles, known as 'medicine bundles'. These were kept by

Opposite *This Indian family posed for a photograph in 1897.*

53

Above *A young Indian 'brave' begins his 'vision quest' in the wild.*

Below *Young warriors used the Sun Dance to fulfil vows by self-inflicted torture.*

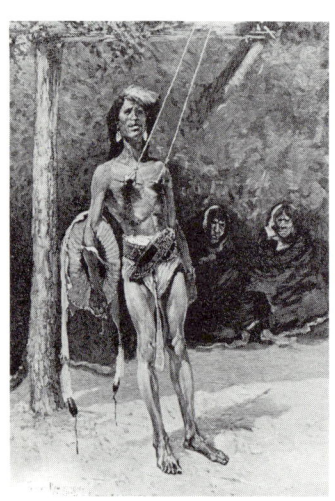

individuals for their own benefit or in trust for the benefit of the tribe as a whole. The sacred bundles were used to enlist the help of spiritual powers in giving them health, wealth and good fortune at all times. Sometimes this involved elaborate ritual ceremonies led by the senior men of the tribe.

An important religious ceremony, the Sun Dance or Medicine Lodge Dance, was held once a year when the tribe gathered for the summer hunt. Although all members of the tribe took part in some way, a few used the ceremony to fulfill vows by self-inflicted torture. The focus of this part of the ceremony was a tree trunk or pole, around which the main participants gathered. In some cases men tied thongs through their flesh which were attached at the other end either to a buffalo skull or to the pole. By pulling against the object they tore themselves free; wounds achieved in this way were a mark of distinction and a source of pride.

War and recreation

Unlike many other societies the Plains Indians did not engage in war to gain territory. Yet they were an aggressive people in the sense that much of their culture centred on the man's role as a courageous hunter and warrior. Although war parties often went out to seek revenge on another tribe, or most often to steal horses, the motive was also to obtain glory and thereby gain status within the tribe. Often the death of an enemy was incidental; a man was considered much more courageous to have done a daring deed and escaped, regardless of whether he had killed or not. However, scalps were taken, the bodies of dead men were mutilated and prisoners sometimes suffered the fate of torture.

Weapons used in warfare, before the Indians obtained the gun, included the bow and arrow, spears and clubs. Shields made from hides were used as protection. On the return to camp victories were celebrated with the men dressed in their war finery, and the women and children performing dances.

Below *The Indians were a warlike people. This drawing shows a traditional scalp dance.*

Children practised for adult life by making games of the activities which would be required of them as they grew older. So boys played at bow and arrow, either shooting at an inanimate object or at small game such as prairie dogs. Both men and boys played the game of snow-snake-propelling darts along the frozen ground to see whose would go the farthest. Racing – both on foot and later on horseback – was a popular sport, and gambling was a popular game among the men.

Disease and guns

Although the re-introduction of the horse had generally worked for the benefit of the Plains Indians, the Europeans introduced other things that eventually led to the destruction of their traditional way of life. Disease was one of these, particularly in the form of smallpox which was responsible for killing thousands of individuals and in some

Above *Sitting Bull, chief of the Sioux Indians, in full ceremonial dress. He led his tribe in one of the last great wars to defend Indian territory from white settlers.*

Right *European settlers selling guns to the Indians.*

cases almost wiping out whole tribes. The fur trade was another. The introduction of the gun enabled the Indians to kill many more buffalo than was necessary for food. In the hands of the Europeans themselves, it resulted in the disappearance of the buffalo herds and many other animals in order to sustain the trade in hides in the eastern states.

The population of the buffalo has been estimated at some 75 million before the arrival of the Europeans. By 1889 their numbers had been reduced to only about 540 individuals. The effect of this on the Plains Indians was disastrous, for it meant that their means of survival was gone completely. In addition, the building of the railroads which opened up areas formerly too remote for settlers increased the demand for land. For governments in Canada and the United States the solution was to confine the Indians to particular areas of land, called reservations, and encourage them to cultivate crops. Although some tribes tried, unsuccessfully, to forcibly resist, others were too dejected and demoralized to protest.

Below *The destruction of the Buffalo and the opening up of the West by the railroads spelt disaster for the Indians.*

GRAND RUSH

FOR THE

INDIAN

TERRITORY !

NOW IS THE CHANCE

TO

PROCURE A HOME

In this Beautiful Country!

Over 15,000,000 Acres of Land

NOW OPEN FOR SETTLEMENT !

Being part of the Land bought by the Government in 1866 from the Indians for the Freedmen.

THE FINEST TIMBER !
THE RICHEST LAND !
THE FINEST WATERED !

WEST OF THE MISSISSIPPI RIVER.

Every person over 21 years of age is entitled to 160 acres, either by pre-emption or homestead, who wishes to settle in the Indian Territory. It is estimated that over Fifty Thousand will move to this Territory in the next ninety days. The Indians are rejoicing to have the whites settle up this country.

The Grand Expedition will Leave Independence May 7, 1879

Independence is situated at the terminus of the Kansas City, Lawrence & Southern Railroad. The citizens of Independence have laid out and made a splendid road to these lands; and they are prepared to furnish emigrants with complete outfits, such as wagons, agricultural implements, dry goods, groceries, lumber and stock. They have also opened an office there for general information to those wishing to go to the Territory. IT COSTS NOTHING TO BECOME A MEMBER OF THIS COLONY.

Persons passing through Kansas City will apply at the office of K. C., L. & S R R, opposite Union Depot, for tickets.

ABOUT THE LANDS.

In answer to inquiries concerning these government lands in the Indian Territory, Col. E. C. Boudinot sends the following from Washington:

FIRST—In reply I will say that the United States, by treaties made in 1866, purchased from Indian tribes, in the Indian Territory, about 14,000,000 acres of land.

SECOND—These lands were bought from the Creeks, Seminoles, Choctaws and Chickasaws, by their treaty of 1866.

The Creeks, by their treaty of 1866, sold to the United States 3,250,560 acres, for the sum of $975,168. The Seminoles, by their treaty of 1866, sold to the United States 2,169,080 acres, for the sum of $325,362.

The Choctaws and Chickasaws, by their treaty of 1866, sold to the United States the "leased lands" lying west of 98 degrees of west longitude, for the sum of $300,000. The number of acres in this tract is not specified in the treaty, but it contains about 7,000,000 acres. (See 14th vol. Statutes at Large, pages 755, 769 and 786.)

Of these ceded lands the United States has since appropriated for the use of the Sac and Foxes 479,667 acres and for the Pottawatomies 575,877 acres, making a total of 1,055,542 acres. These Indians occupy these lands by virtue of treaties and acts of Congress. By an unratified agreement, the Wichita Indians are now occupying 743,610 acres of these ceded lands. I presume some action will be taken by the United States government to permanently locate the Wichitas upon the land they now occupy. The title, however, to these lands is still in the United States.

By executive order, Kiowa, Comanche, Arrapahoe, and other wild Indians, have been brought upon a portion of the ceded lands, but such lands are a part of the public domain of the United States, and have all been surveyed and sectionized.

A portion of these 14,000,000 acres of land, however, has not been appropriated by the United States for the use of other Indians and all probability never will be.

THIRD.—These unappropriated lands are situated immediately west of the 97th degree of west longitude and south of the Cherokee territory. The soil is well adapted for the production of corn, wheat and other cereals. Is is unsurpassed for grazing, and is well watered and timbered.

FOURTH.—The United States have an absolute and unembarrassed title to every acre of these 14,000,000 acres, unless it be to the 1,054,544 now occupied by the Sac and Fox and Pottawatomie Indians. The Indian title has been extinguished. The articles of the treaties with the Creeks and Seminoles, by which they sold their lands, begin with the statement that the lands are ceded "in compliance with the desire of the United States to locate other Indians and freedmen thereon." By the express terms of these treaties the lands bought by the United States were not intended for the exclusive use of other Indians" as has been so often asserted. They were bought as much for the negroes of the country as for Indians.

ADDRESS

WM. C. BRANHAM,

Independence, Kansas.

To parties accompanying my Colony, I would advise them to purchase their Outfit at Independence, Kas., I have examined Stock and Prices of Goods, such as Wagons, Plows, Lumber, Dry Goods, Groceries, and, in fact, everything that is needed by Parties settling upon new Land, and find them as cheap as they can be bought in the East.

RESPECTFULLY YOURS,

Col. C. C. CARPENTER.

P. S.—Parties will have no trouble in getting transportation at Independence for hauling their goods into the Territory.

C. C. C.

Left *Advertisements like this encouraged white settlers to take over Indian lands.*

Above *Indians today in Monument Valley, Utah.*

7 Shepherds of today—pastoral tribes in East Africa

Modern farms, which produce food on a large scale for the commercial market, exist in East Africa as they do in other parts of the world. Yet on the grasslands of Kenya, Uganda and Tanzania, many tribal peoples continue to lead a way of life that they have followed for thousands of years. Despite pressures from governments, land developers and settlers, they have managed to preserve a traditional and very distinctive life-style.

The pastoral tribes of East Africa include the Masai of Kenya and Tanzania, the Karamajong of Uganda and the Dinka or Nuer of the Sudan. All belong to the same large group of peoples who orginated further north along the River Nile. Today the majority of Masai live on reserves in Kenya–on land ranging from bush to fertile grassland–and on the savannahs and in the highlands of Tanzania. The Karamajong occupy lands in north-eastern Uganda, in central Karamoja. As with most African peoples, much of their early history is unknown; the introduction of writing came only recently to regions south of the Sahara, and the traditions and cultures of these peoples were therefore passed on verbally.

Generally pastoral tribes live in areas of low rainfall, which makes the land unsuitable for the permanent growth of crops. However some of the lands occupied by the Masai are fertile enough for cultivation. In the past, competition for land led to conflict with farming peoples, and raiding parties earned the Masai the reputation of being fierce and warlike. Today they live fairly peaceably with their neighbours but their attitude to the land is unchanged. To the Masai, land is free for those who wish to wander over it with their herds. These peoples are therefore hostile to farmers, who destroy the grass and pastureland when they cultivate their crops.

Practically every aspect of the pastoral peoples' lives is

Opposite *Masai warriors in ritual dance.*

Below *A Karamajong mother and child from north-eastern Uganda.*

61

Above *A Masai child herding cattle. These people grow up with an immense pride in their animals.*

centred around the animals that are their main source of livelihood. Cattle, mainly zebu cattle, are their most prized possessions, but they also herd sheep and goats, while donkeys or asses may be used for transport. Camels are sometimes kept for both transport and as a source of food. The Turkana peoples, however, milk and eat the meat of camels, but keep them to herd rather than ride.

The need to move from one place to another in search of pastures and water suggests that these pastoral people lead a nomadic life, but in reality they usually have semi-permanent settlements to which they return at different seasons, and it is the young men who usually herd the animals to and fro. Often their movements are between dry and wet season camps; during the dry season they congregate in large numbers at water sources such as rivers, lakes and water holes, while in the rainy season they disperse throughout the pastureland.

Cattle, sheep, goats, and camels provide the pastoral peoples with milk and blood, both of which make up a major part of their diet. The blood is obtained by tying a thong about the animal's neck and tightening it just

Above *A semi-permanent Masai village on the banks of a river.*

enough to swell out the jugular vein. A sharp tool is used to prick the vein, from which blood is then drawn. This operation is painless to the animal and the small puncture usually closes over quickly, although sometimes clay may be worked into the wound to block it and also help prevent infection. As each animal can be bled safely only once every two months, six or seven animals per person are needed to provide a family with an adequate diet. Animals may be hunted occasionally to add meat to the diet, but for some peoples, such as the Masai, the eating of meat is restricted to ritual sacrifices.

Pastoral peoples gather berries and fruit and grow some crops. These will be cultivated after the rainy season, or at other times of the year where soils and rainfall are favourable. Such crops are usually sorghum and maize. Sorghum is a native plant of Africa; its main advantage is that it can withstand drought and its wilted leaves revive when the rains come. It also ripens quickly and can be harvested in about two months. Maize, which was introduced from North America, demands a richer soil and heavier rainfall, but it gives high yields. Apart from

Above *Moving home with laden donkeys is a regular occurrence for the wandering Turkana people.*

occasionally raising a few crops, very little attention is given to the management of the land. However, grass and bush are sometimes burned to improve grazing and to reduce the number of insect pests.

The family is the basic social unit of pastoral people, but communities may also be divided into different groups according to age. Each age group has certain duties which they must perform for the benefit of the family or tribe. Among the Masai, for example, young men between certain ages are known as *moran*. The moran are responsible for protecting other members of the community and the herds, but they are also expected to help with any task that needs doing, however menial. Traditionally, they also carried out raids on neighbouring tribes, particularly the farming Kikuyu.

When a man marries, his family give presents of cattle, called bridewealth, to his wife's family. Men may have more than one wife, and each is given certain cattle to care for. The members of the community live in family groups in a settlement called an *enkang*. These are usually made up of several mud huts plastered with dung within a fortified wall of thorny branches. The herds are taken out to pasture every day, but at night they are protected from raiders within the enkang. The Masai moran are usually separated from other community members and live together in settlements called *manyatta*, from where they can watch over the enkang. These young men go to great trouble to beautify themselves; they paint their bodies with ochre and braid their hair, often decorating it with beads.

The Masai's religious beliefs centre around their cattle and the daily activities of caring for them. *Ngai* is the name of the creator and the giver of cattle. Cattle are not only a symbol of wealth and power; they also represent the qualities of goodness and beauty. Songs and dances are created and performed in their honour; certain cattle are given names and their bodies may be decorated and even painted. This concern with cattle is even expressed in the rites for the dead. The body of the dead one is placed in the open. It is now a useless thing to be disposed of by scavengers. The spirit however is given materials that it will need in the afterlife: grass in one hand, and a pair of sandals and a stick for herding in the other.

8 Farmers and ranchers

The main object of the pastoral tribesman is to produce enough food for his family to live on. His herds and the crops he grows are his only means of livelihood, and there is little if anything left over for trading. Such a way of life is known as *subsistence agriculture*. It is practised by a large number of the world's peoples, not only by those who herd cattle but also by those whose main occupation is the cultivation of crops.

On the dry, savannah grasslands of India the main agricultural products are rice, wheat and millet, and the majority of farmers who raise these crops are part of a subsistence economy. In many areas the only agricultural implement is the wooden plough, drawn by an ox, and there are still places that are cultivated solely with a hoe. The farmers work small areas of land (more than half of them cultivate less than three hectares). Often they do not have proper irrigation, which means they are dependent on the monsoon to bring the much-needed rains to water the fields. Failure of the rains to arrive brings drought followed by famine. Too heavy a rainfall results in flooding, and where overgrazing and overcultivation have taken place, erosion of the already depleted soils. Events such as these are disastrous for farmers anywhere, but especially in India where most of them cannot grow sufficient food to feed their families even in good times.

At the opposite extreme to the subsistence farmer is the modern rancher or farmer who consumes little or nothing of the meat or crops he produces. The produce of his farm is transported to towns and cities to be sold on the national or international market. Commercial agriculture is confined largely to the temperate grasslands, most of which occur in the high income, industrialized countries such as the United States, Canada, Russia, Australia, and New Zealand. Exceptions to this are generally found in tropical and sub-tropical regions where Europeans settled and brought modern methods of agriculture with them.

Opposite *In many parts of India the only agricultural implement is the wooden plough, drawn by oxen.*

H E R E   AND   T H E R E ;
Or, Emigration a Remedy.

Above This nineteenth-century cartoon contrasts poverty in Britain with the good life beckoning from the plains of the New World.

Opposite An artist's impression of pioneer life on the American prairies.

The grasslands of the New World are largely owned by peoples of European descent. The major period of settlement, during the nineteenth and early twentieth century, brought immigrants from a number of European countries. Many came from areas where land was scarce; they were induced by offers of free or cheap land in return for the promise to settle for a fixed period and improve their homestead. Others came to escape religious or racial persecution in their native countries. The earliest of the Australian settlers were convicts from Britain, who had been deported to provide free labour in the developing colonies. But later they were followed by immigrants who, like those in the Americas, came in search of a new and better life.

The immigrants who chose to open up the prairies, pampas and plains of the New World were accompanied by a greater number of people who chose to make their lives in the towns and cities. The growing urban populations

On the Road

Crossing a River

The Second Season

The First Season

WR

needed increasing amounts of food and it was this need, plus a general world demand for food, that resulted in governments encouraging settlers to pioneer the development of the interior grasslands by raising livestock or cultivating crops.

The early pioneers worked their land with an iron plough pulled by horses; such work was difficult on the heavy, fertile soils that cover many of the temperate grasslands. With the invention of the steel plough in the 1830s they were able to cultivate more land at a more efficient rate. (This invention also had a similar effect on the steppelands of Eurasia.) However it was the development of agricultural machinery in the latter half of the nineteenth century that brought an even greater area of land under cultivation. Machines for reaping and threshing, steampowered ploughs and tractors were in use by the 1890s. The combine harvester was invented during this period, and although for a time it was drawn by a traction engine it

Below *A very early photograph of a pioneer family on their way out West.*

was not long before it operated under its own power, driven by an internal combustion engine.

The introduction of agricultural machinery meant that fewer farm labourers were required to help work the land, and as its use became widespread there was a gradual drift from the country to the towns. As a result, the modern farmer is part of a highly complex agricultural system, in which food for many is produced by a few. The production of food crops on a large scale is achieved by the use of fertilizers, pesticides, the control of water systems, and the breeding of highly productive, disease-resistant crops, as well as the use of agricultural machinery.

The main crops grown by farmers on the grasslands are cereals–highly refined grasses whose seeds, or grains, are the basic source of food for most of the world's peoples. Wheat is the cereal best adapted to the temperate conditions found in Australia, Canada, Argentina, the United States, and Russia, and is one of the principal crops of these countries. It is grown in areas that have moist, moderate climates in spring to promote growth, and warm, dry summers needed for harvesting. Most of the wheat production is used to manufacture bread, while maize, another important cereal crop, provides food for both people and domestic animals. Maize requires hotter summers and more moisture than wheat and is therefore grown in areas that have either reliable summer rains or adequate irrigation systems. It is a major product of the Argentinian pampas and the North American prairies.

Rye is one of the hardiest cereal crops and grows well in moist, cool regions of temperate grasslands, especially those in eastern Europe and Russia. Both rye and oats, another cereal of cool areas, are usually fed to livestock but they are also eaten as breakfast foods and breads. They make up a large proportion of the grains produced in the Ukraine, the richest cereal-producing region of the Soviet Union.

The Ukranians cultivate the black, fertile soil of the western steppe, living and working on collective farms (*kolkhoz*). The kolkhoz is often organized around several

Above *Huge grain elevators dominate the main routes across the North American prairies.*

Above *A line of combine harvesters works across the rolling wheat fields of the prairies.*

villages, and the state-owned land is leased to the farmers of the collective. They are paid according to the type of work they do, and the amount of food they produce. On state farms (*sovkhoz*), the farmers work directly for the state and are paid a fixed wage.

Areas of grassland that are unsuitable for growing crops (due to low rainfall and less fertile soils) are used as pastureland for livestock. Ranchers in Argentina raise beef cattle on the huge estates of the pampas; in Australia cattle are bred on 'stations' or 'properties' on the savannah grasslands of Queensland and the Northern Territory. In both countries ranches may cover tens of thousands of hectares and are often worked by hired hands (*gauchos* in South America, *jackaroos* in Australia) under the direction of the owner or manager.

Early Australian farmers brought Hereford cattle into the country but these animals could not adapt to the heat and drought. A hardier breed was produced by crossing them with zebu cattle. The result was a type that could cope with the dry, arid conditions and was also resistant to pests such as the cattle tick. The herds of beef cattle on the prairies (originally Longhorns of Texas) and pampas were similarly cross-bred with pure breeds from Europe to produce cattle that gave high quality meat.

Dairy cattle require much wetter conditions than beef cattle. They therefore do well on temperate grasslands in North America, south-eastern Australia and on North Island in New Zealand.

Sheep as well as cattle are an important resource on temperate grasslands throughout the world, and are raised for wool and meat. Merino sheep, originally bred in Spain during the fourteenth century, are a hardy breed that produce fine wool. Ranchers in New Zealand crossed them with an English breed to produce the Corriedale, a type that gives both good meat and wool.

The cultivation of high-grade cereal grasses to suit the various climatic conditions of grasslands includes raising grasses and other plants that are sown as pastureland.

Below Rounding up cattle in Arizona, USA.

Alfalfa (or lucerne) is grown as a pasture grass on many grasslands. It is a *legume*, a member of the pea family, and has a penetrating root system that enables it to absorb water from deep within the ground. When other grasses naturally wither, it continues to flourish. Like other legumes its growth enriches the soil with nitrogen. Crops that result from one sowing will provide cattle with fodder for up to ten years.

In Australia and New Zealand large areas of land have been cleared of 'bush' and sown with cultivated grasses which provide excellent pastureland. This increases both the quality of grazing and the amount of grazing land available for livestock.

Although the main use of grasslands has always been agricultural, industry has developed on plains and prairies where there are rich deposits of minerals and fuels. For example gold was mined for hundreds of years by the native peoples of southern Africa, and today gold and uranium are mined on part of the South African veldt. The North American prairies yield substantial amounts of natural gas, oil and coal.

Left *Huge sheep stations exist on the grasslands of Australia.*

9 Man's mismanagement of grasslands

In the spring of 1934 the Great Plains of North America were the scene of a disaster that put thousands of farmers out of work and destroyed millions of hectares of land. Tornadoes, moving eastwards across the states of Texas, Oklahoma, Kansas, and eastern Colorado, swept up the dry dusty soils that covered the land and carried them towards the Atlantic coast. In some places 25 centimetres (10 in.) of surface soil were removed by the winds, exposing the coarser, infertile soils that lay beneath them. Measures taken to restore the grassland have been only partially successful, and large areas of man-made desert, known as the Great American Dust Bowl, still remain where once there was productive land.

The creation of this desert in America is an extreme example of Man's destruction of grasslands, but it illustrates how bad management can result in the loss of valuable habitats. For several decades before the disaster struck, the Great Plains had been continuously cultivated with cereal crops. These robbed the soil of valuable nutrients that were not replaced by the farmers, and the result was to break up the soil texture. Between plantings, the land was left exposed or cattle were allowed to graze the stubble left after harvest, thus continuing the degrading of the soil structure by weather and trampling. When in 1930 a ten-year drought began, the soil turned to dust. Similar events occurred on other parts of the prairie, including an area of Alberta, and the Great Plains themselves suffered from soil erosion again in later years.

A certain amount of erosion occurs naturally, but the risk of soil erosion by wind or rain increases when the natural plant cover is removed by ploughing and is replaced by shallow-rooting crops. It is also accelerated by allowing livestock to graze too long in one area, or putting too many domestic animals on the land. Overgrazing by livestock has been a problem on most grasslands but it has

Opposite Overgrazing has resulted in the desert encroaching at an alarming rate in East Africa. A mother and child stand in the sand which covers their village.

77

been most severe on the prairies and on the African savannahs.

In North America, as elsewhere, livestock graze lands that are too dry for crops. Vegetation is therefore sparser in these regions and takes longer to grow. When cattle and sheep are put to pasture in large numbers and are restricted to one area, they destroy most of the vegetation. This is because, unlike wild herbivores, they graze plants right down to the roots. As the plants die and are not replaced the soil is exposed and is vulnerable to erosion. Large areas of grasslands have been lost in this way through the carelessness of ranchers, particularly during the last decades of the nineteenth century.

A similar situation exists on the grasslands inhabited by the pastoral peoples of Africa. Prior to the arrival of the Europeans, there was less risk of overgrazing because the shepherds could wander with their herds over a wide area in search of fresh pastures. Also, the presence of disease-carrying insects such as the tsetse fly acted as a natural control on the number of livestock, as did the cattle disease, rinderpest. Since the movements of the pastoral peoples have been restricted however, and rinderpest has been largely controlled by vaccinating cattle, overgrazing has been a major problem. This has been aggravated by the natural inclination of these peoples to own as many cattle as possible. Also, poor grazing results in the inability of cattle to give adequate amounts of milk and blood, so more cattle are needed to provide sufficient supplies. The result is that the desert is encroaching at an alarming rate in some parts of East Africa.

In India, trampling and gullying of the land, as well as destruction of vegetation, occurs wherever the sacred cows of the Hindus graze.

The breeding and raising of domestic animals can have another serious effect on grasslands. It can result in the decrease, and sometimes the extinction, of wild animals. The domestication of animals itself has resulted in the disappearance of most of the wild animals from which they were originally bred. In addition, domestic animals

Below *These acacia trees have been devastated by a herd of elephants. Wild animals have less and less space to wander.*

compete directly with wild species for food. Where they graze, there may be little or no food left for others, who must either find alternative sources, or starve. Often, the wild animals are killed by farmers and ranchers who consider them 'pests'. This happens in Australia where kangaroos have been slaughtered by the millions (and are now in danger of extinction) in the belief that they feed on resources needed for cattle and sheep. In fact, while kangaroos can sometimes compete locally with livestock, generally they survive on those plants that are rejected by the more selective domestic animals.

Wild animals that inhabit grasslands are also killed by Man through ploughing up the land, which destroys their habitats, and by the use of pesticides and fires. These factors also work to destroy the natural variety of grassland vegetation. Fire, for example, promotes the growth and spread of fire-resistant plants, but it destroys the more fragile species. Selective chemical sprays, which destroy 'weeds'–often the most interesting of the wild flowers– also reduce the variety in the plant community, and consequently the number of animal species that can live there.

Below *Fire destroys many fragile plant species. A grass fire at night on the African savannah.*

10 What of tomorrow?

He gave it for his opinion, that whoever could make two
ears of corn or two blades of grass to grow upon a spot of
ground where only one grew before, would deserve
better of mankind, and do more service to his country
than the whole race of politicians put together.
(Jonathan Swift, *Gulliver's Travels*)

The question of how we should best manage the grasslands
in the future involves two fundamental problems. The first
is that there are 4,000 million people in the world and
some two-thirds of them live in poverty. Many of these
suffer all their lives from chronic hunger and severe
malnutrition. The second problem is the need to conserve
the remaining natural grassland habitats and the animals
that live on them, and the need to use sound agricultural
methods on the cultivated grasslands to prevent further
destruction.

Feeding the world's peoples

The problem of providing an adequate diet for the starving
is a vast and complex one. It is not merely a matter of
growing more food in those countries which are fortunate
enough to have the benefit of fertile soils, and modern agricultural equipment and methods. The world's economy is
so organized that there is an imbalance in the way food-
stuffs are distributed between nations. This becomes
obvious when European nations (among others) accumu-
late wheat, apple, butter, and other food 'mountains'
which are then ploughed back into the land or dumped into
the sea. Neither is it possible, in a world where great social,
racial and economic divisions exist between peoples (both
within and between nations), to ensure that raw materials
and technology will benefit those who need them most.

Scientists have attempted to raise agricultural production by improving the strains of wheat and rice

Opposite *It has been
estimated that 75 million
buffalo lived on the
American prairies before
the arrival of the
Europeans. Today only a
few herds survive.*

81

Above *Famine is on the increase in the Third World. Starving boys in northern Uganda.*

cultivated by local farmers in the poorer countries. The initial results in the 1960s, with the new seeds doubling and even tripling yields, gave rise to great optimism—it seemed as though the 'Green Revolution', as the project came to be called, had arrived. There were, however, certain problems connected with the new plants. Very high levels of fertilizer, pesticide and water were required, and if any of these were lacking, then the resulting yield would be less than for the crops the new seeds had replaced. In addition, the 'miracle' seeds were found to be much less resistant to disease, and any blight that appeared would quickly sweep through the entire crop. In fact the project really only benefited the medium to large farmer who could afford to buy the fertilizer and modern machinery the new seeds required. The rural poor could not compete, and were, if anything, worse off than before.

Similarly, attempts to 'transplant' agricultural methods from one area to another are often unsuccessful. The combine harvester, which was an efficient aid in raising wheat production on grasslands in the New World, is not necessarily the answer to the problems of farmers in other parts of the world. It is a costly item that is only used for a very short period in the year and is only worthwhile when used on large areas of land and is shared between farmers.

The solution to the food shortage may be provided by individual countries who adopt those methods which are best suited to their own circumstances. In Japan, land reforms since the Second World War have largely put an end to huge estates and have returned the ownership of the land to those who work it. There are now more opportunities for employing farm workers and the yield per hectare has risen remarkably. It is interesting to note that this has been accomplished by the policy of having small farms, and by using only small machines such as tractors.

As the population explosion continues, our present food resources will never be able to provide enough food for everyone, even if all the inequalities disappear. A partial solution may lie in developing and exploiting new types of food such as algae and plankton. The choice of foods may also have to be restricted with changes in farm practices. The production of meat from domestic animals, for example, uses a large proportion of land that could be converted for crops or used to graze 'tamed' wild herbivores that use only one tenth of the amount of grazing land, and are resistant to a large number of diseases and pests that affect domestic animals.

Below *The Green Revolution, combining high-yielding seeds and fertilizers, can double rice production in Asia. Unfortunately most small farmers do not have the money or knowledge to introduce the new techniques.*

Food or wildlife?

The landscape has always been a source of inspiration for Man and this is no less true of the grasslands than it is for any other natural habitat. Painters, novelists and poets have expressed the way it affects them and others, in works that are now considered masterpieces. The American painter Andrew Wyath captured the loneliness and desolation of the prairies in his pictures and John Steinbeck wrote of life on the Great Plains during the depression and the creation of the Dust Bowl in *The Grapes of Wrath*. Many of Thomas Hardy's novels are concerned with farming communities and peoples in south England, and British poets have for centuries used rural imagery to develop their themes.

The world's grasslands are worth conserving not only as productive land but as land that provides something in addition to bread. Unfortunately the interests of farmers and conservationists sometimes clash over the use land should be put to, although there are many farmers who are also conservationists. Farmers are constantly urged to produce more food and in some cases they respond by clearing more and more land where woodland, hedgerows, marshland, and moors have existed for years. In Britain, where most of the land was once covered by broadleaved forest, most of it is now under cultivation (some 80 per cent is classified as agricultural land) either for crops or as grazing for domestic animals. As a result, plants and wild animals have disappeared from the landscape and are continuing to do so in larger numbers every year. The problem of land use has been expressed as a question of choice–food or wildlife? But it may be possible to find a solution without having to decide between the two, even if it is a compromise.

Opposite *Aid organizations provide soup for malnourished children in Kenya.*

Glossary

Anthropologist   A person who studies mankind, especially his physical characteristics, culture and social behaviour.

Archaeologist   A person who studies the life and cultures of ancient peoples by examining ancient sites and the fossils and other objects found there.

Bush   A term used in Australia and New Zealand to describe woodland and forest.

Campos   Tropical grassland of central and south-eastern South America.

Carnivore   An animal that eats other animals. A lion is a carnivore.

Conservation   The protection of natural resources, including the environment, plants and animals.

Ecosystem   A plant and animal community and the environment they share, all of which interact with one another.

Forbs   All non-woody plants other than grasses.

Grazer   Any animal, but particularly a mammal, that feeds on grasses and herbs.

Habitat   The environment, including vegetation, in which an animal lives.

Herbivore   An animal that eats plants rather than the flesh of other animals.

Legume   A plant belonging to the pea family, such as beans, clovers and vetches.

Llanos   Tropical grassland of northern South America in the Orinoco River Basin of Venezuela.

Longitudes   Imaginary lines on the Earth's surface that run north and south between the two poles. They are used to measure distances east and west of Greenwich in Britain.

Monsoon   The prevailing winds of South Asia which blow from the south-west in summer (wet), and north-east in winter (dry).

Nomads   Wandering people who frequently change their habitation to find food for themselves and their livestock.

Nutrients   Any of a number of chemical compounds that provide nourishment for growth and development.

Omnivore   An animal that eats both plants and animals. Man is an omnivore.

Pampas   A Spanish word for plain, used to describe the

temperate grasslands of South America.

Pastoralism   The herding of sheep, goats, cattle and other domestic animals as a way of life.

Prairie   The large belt of temperate grassland in central North America; it is the French word for 'meadow' and was used by early French explorers when they first saw the grasslands.

Predators   Animals that hunt and kill other animals.

Rhizomes   Plant stems that grow underground.

Rodent   A group of small vegetarian mammals whose teeth, particularly the incisors, are adapted for grinding foods; the word rodent means 'gnawing animal'.

Savannah   A general term used to describe tropical grasslands that are characterized by scattered trees and shrubs; often used specifically for the tropical grasslands of Africa.

Scrubland   Land that is dominated by shrubs and low trees; generally found in areas where rainfall is too low to promote the development of woods or forest.

Steppe   Temperate grassland found in areas of low rainfall; used especially to describe the grasslands of eastern Europe and Asia (the word is Russian for plain).

Stolens   Plant stems that creep along the ground.

Subsistence agriculture   A method of farming which predominates in many poor countries, where the farmer produces only enough food for his family to live on, with little or nothing left over for trading.

Veldt   Temperate grassland of South Africa.

Book list

BAINES, JOHN D., *The Environment* (B. T. Batsford: 1973)

CHRISTIANSEN, M. SKYTTE, *Grasses, Sedges and Rushes* (Blandford Press: 1977)

CLOUDSLEY-THOMPSON, JOHN and DUFFEY, ERIC, *Deserts and Grasslands: the World's Open Spaces* (Aldus Books/Jupiter Books: 1975)

CUISIN, MICHEL, *Nature's Hidden World: Animals of the African Plains* editor, CHINERY, MICHAEL (Kingfisher Books, Ward Lock Ltd: 1980)

HORTON, CATHERINE, *A Closer Look at Grasslands* (Hamish Hamilton: 1979)

MOSCATI, SABATINO, *Archaeology* (Collins Publishers/Franklin Watts Inc.: 1975)

ROGERS, EDWARD S., *Indians of the Plains* (The Royal Ontario Museum: 1970)

RUSSELL, W. M. S., *Man, Nature and History* (Aldus Books: 1967)

SELLMAN, R. R., *The Prairies* (Methuen: 1974)

SIKULA, DR JAROMÍR, *Grasses* (Hamlyn: 1978)

STEINBECK, JOHN, *The Grapes of Wrath* (Heinemann: 1939)

TURNER, GEOFFREY, *Indians of North America* (Blandford Press: 1979)

WHITEHOUSE, RUTH, *Your Book of Archaeology* (Faber & Faber: 1979)

Picture acknowledgements

Biofotos—Heather Angel 18, 23, 26; Bruce Coleman front cover, 6, 9, 11, 14, 20, 21, 33, 35, 62, 63, 71, 74, 79; Colourpix 67; Bill Donohoe 12, 43; Mary Evans 8, 36; FAO 84; Ian Griffiths 19, 28 (below), 30, 64; Alan Hutchison frontispiece, 15, 45; Mansell Collection 38 (above), 39, 57, 68, 69; NHPA 24, 25, 28 (above), 29, 31, 32, 49, 73, 78, 80; Oxfam 10, 41, 61, 76, 82; Vision International 60; Wayland Picture Library 34, 38 (below), 42, 44, 51, 55, 70, 83; Western Americana 46, 50, 54 (both), 58, 72; ZEFA 17, 52, 59.

Index